AF604771

Australian
GEOGRAPHIC

CREATURA

Processionary caterpillar

Australian Geographic

CREATURA

Weird and wonderful Australian animals

Bec Crew

CONTENTS

Pig-nosed turtle

BIRDS

REPTILES AND AMPHIBIANS

INTRODUCTION

When the Creatura Blog launched back in 2013, I thought to myself, 'How on Earth am I going to find a different animal each week that's strange enough, obscure enough, and downright charming enough to celebrate with the Australian Geographic community?'

But here we are, eight years later, and the pipeline of peculiar is anything but dry. I mean, have you seen the head-stacking caterpillar?

What I've learned from my time writing about weird animals is that there is always something weirder than you could have ever imagined just around the corner. From bright blue bees to furry little squat lobsters and all manner of hot-headed marsupials, the skies and the seas and everything in-between are full of these strange critters – you just have to know where to look.

The Creatura Blog highlights animals that hail from Australia and its closest neighbours in the Oceania and South East Asia regions. This book is a selection of the Australian contingent specifically, and runs the gamut of mammals, birds, reptiles and amphibians, invertebrates, and fish and sea creatures.

What stands out in many of the Australian species featured in this book is an unmistakable hardiness. The ningaui, for instance, is tiny in stature, but has the ferocity of an animal 10 times its size. These rambunctious desert-dwelling marsupials are notorious biters, as I'm sure any researcher who has handled one can attest.

The desert spadefoot, another species that has adapted to life in a hostile environment, spends most of its days buried in the sand, only emerging after rainfall to feed, mate, and lay eggs in an urgent flurry of activity.

Ingenuity is another quality that Australian animals have in, well, spades. The whistling kite is a wonderful example. For thousands of years, it's been extending the range of bushfires using lit twigs like candles to flush prey out into the open. The banded archerfish has figured out how to target prey 2 or 3 metres above the surface by spitting powerful jets of water at them. And fishing spiders? Let's just say Australia is a record-breaker in that department, and I'm not sure how I feel about it.

This country is renowned for its dangerous animals. Even those that aren't dangerous, like the daddy longlegs, find themselves embroiled in myth and legend to make them seem more intimidating than they actually are. Our animals are also stunningly beautiful. The blanket octopus, with its sheets of flesh trailing behind it like technicolour curtains, is absolutely magnificent. And the juvenile dusky batfish is positively otherworldly, its pitch-black body rimmed in fiery gold.

If I had to choose one creature in this collection that stands out among the rest, it would be the southern marsupial mole. With a luxurious golden coat, huge claws, a nubby tail, and no eyes, it's like someone plucked it from a witch's cottage in *Grimm's Fairy Tales* and plonked it onto the red desert sands of central Australia.

It's hard to imagine an animal that's stranger and more charismatic than this wonderfully grotesque character, but I'm sure it's out there, lurking on our doorstop.

Flamboyant cuttlefish

Common spotted cuscus

MAMMALS

NINGAUI

The chihuahua of the Australian desert, the ningaui will mess you up if you so much as look at it the wrong way. That might be an exaggeration, but you don't want to cross a ningaui if you enjoy having all of your fingers.

Carnivorous marsupials with a taste for insects, spiders and lizards, ningauis look like dunnarts, but can be told apart by their broader hind feet and more bristly, messy-looking fur.

They are nocturnal and elusive, which, along with their tiny bodies, makes them tough to spot. They also have the perfect cover – the hummock (or spinifex) grasses of inland Australia are just the thing to dart in and out of when being pursued by larger creatures.

It's not difficult to be a larger creature. The smallest of the three ningaui species, the Pilbara ningaui (*Ningaui timealeyi*) is just 5.8cm long, with a tail of just over 7cm. This makes it almost half the size of a house mouse. In fact, the Pilbara ningaui is one of the smallest marsupials on Earth, alongside the minute planigales, also of Western Australia.

The Pilbara ningaui also has the narrowest range of the three – it's found in the Pilbara and Gascoyne regions of Western Australia and also out into the Little Sandy Desert.

The southern ningaui (*Ningaui yvonneae*) is found in northern Australia, and in pockets in Western Australia, South Australia and Victoria. Slightly larger than the Pilbara ningaui, it can grow to 7.4cm long, with a tail of about 7cm. It's distinguished from its relatives by its greyish, olive colouring.

The Wongai ningaui is ever so slightly larger than the southern ningaui, and has the widest range of them all, from the west of Kalgoorlie in Western Australia to northern South Australia, then up in the southern parts of the Northern Territory and over to south-western Queensland.

The name ningaui refers to an Aboriginal legend that describes imp-like creatures that live in the mangroves and lure passers-by to their deaths, consuming them raw. While that might be more diabolical than what the ningaui marsupials are capable of, it's not for a lack of trying.

CHRISTMAS ISLAND FLYING FOX

This rare Australian bat is an unusual sunseeker. We've a lot of love for this daylight-loving larrikin.

Bats and the night are like house cats and naps – two concepts so intrinsically linked, it's almost impossible to imagine a sunlight-loving bat or a house cat that doesn't spend its life luxuriating in comfort while the rest of us run around all day like suckers.

But, as it turns out, not all bats are nocturnal. The Christmas Island flying fox does most of its flying, feeding and socialising by the light of day.

Its busiest times are between 2:30pm and 6:30pm, and researchers think it's because it allows them to feed on open flowers and access daily wind patterns to travel between foraging areas.

Christmas Island's only native mammal, this flying fox hangs out in large groups called camps, which can be several hundred members strong. They are important pollinators on the island, feeding on the fruits and flowers of dozens of plant species, including the Japanese cherry (*Muntingia calabura*) and tropical almond (*Terminalia catappa*).

These bats have an interesting lineage – they are either considered an endemic species in their own right (*Pteropus natalis*) or a distinct subspecies of the black-eared flying-fox (*Pteropus melanotus natalis*), another diurnal (opposite of nocturnal) bat.

Sadly, the Christmas Island flying fox is considered Critically Endangered, and has experienced a steep decline in its population in recent years. Back in the early 1900s, researchers described them as common, but by 1984, there were just 6000 of them, confined to the 135sq.km island.

Between 1984 and 2006 the population declined by as much as 75 per cent, and has continued to drop ever since due to hunting, habitat loss, and predation.

Christmas Island flying fox

Cuvier's beaked whale

CUVIER'S BEAKED WHALE

The world record for the deepest and longest dive performed by any mammal has been smashed by a Cuvier's beaked whale off the coast of southern California.

Shaped like a cigar with a short beak and body length of up to 7m, Cuvier's beaked whales (*Ziphius cavirostris*) have tiny flippers and collapsible lungs to help them plunge quickly and safely into the depths of the ocean in search of food.

They're very broadly distributed, found in temperate and tropical waters all over the world, including off the coast of southern Australia, but they're also skittish and shy. Their preference for deep-sea habitats makes them notoriously difficult to study.

In 2014 a team from the Scripps Institution of Oceanography in La Jolla, California announced they had managed to record over 3000 hours of diving activity by Cuvier's beaked whales, and what they found was pretty extraordinary.

The team attached tracking tags to eight Cuvier's beaked whales caught just off San Nicholas Island in southern California and followed them for several months.

During this time, the whales would usually dive to about 1400m below the surface, but one individual decided to keep going, all the way to a depth of 2992m.

This not only exceeded the previous deep-diving record for the species by almost 2km, it also smashed the world record held by a southern elephant seal that clocked a 2388m-dive in 2010.

This exceptional whale also managed to stay down there for 138 minutes, which secured it a second record for the longest dive ever performed by a mammal.

Other than gaining an unprecedented insight into the species' remarkable diving abilities, this study was part of an ongoing investigation into how military sonar operations in the area could be affecting the whales.

The team tried to identify incidents of sonar use that occurred while they were recording the whales to determine if and how this acoustic disturbance was having an impact on these very special animals.

Common spotted cuscus

COMMON SPOTTED CUSCUS

*What a sweet little face. This is the common spotted cuscus (*Spilocuscus maculatus*) from the far north region of Queensland's Cape York Peninsula and all over Papua New Guinea.*

The size of a house cat, and with colours reminiscent of a white chocolate and caramel puff, this species is wrapped in an incredibly beautiful coat. The females have an overall creamy colour with a ginger face and the males are ginger with a network of white splotches arranged like dappled sunlight. This striking coat is what makes the spotted cuscus one of the world's prettiest marsupials, but it also makes it stand out like a beacon in the forest canopy.

With nothing but claws, teeth and some kicky back legs to defend itself, the cuscus is easy prey for large birds like the Papuan eagle, tree pythons, and yes, humans.

One strategy to hide itself is a nocturnal lifestyle, but how does it keep out of sight in the light of day?

In 2002, ecologist Thomas Heinsohn from the Australian National University visited New Ireland, a large island off the east coast of Papua New Guinea. Here he met an adult male spotted cuscus sleeping in a coconut palm.

Cuscuses don't curl themselves up in tree hollows like other possums do. Heinsohn found one resting between the branches, about 12m up, with its head tucked between its legs, in typical cuscus sleeping fashion. Curiously, it looked as if a palm frond had been deliberately pulled down and tucked around its body, obscuring the brightest shades of its coat.

"During steady observation through binoculars over the next hour, the resting cuscus was observed to actively grab hanging palm frond pinnae with its forepaws, pull them in towards itself, and tuck them in around its body to improve its quite effective camouflaging shield of vegetation," Heinsohn reported in the journal *Australian Mammalogy*. "Eventually only patches of its bright white and ginger-red spotted pelage were visible with the aid of binoculars through gaps in the camouflaging foliage." This adorable golden possum sleeps in trees with its head between its legs, wrapped in leaves like a cool green blanket.

STRIPED POSSUM

With its remarkable coat and strange little fingers, the striped possum is a peculiar fellow that few Australians are lucky enough to see.

The striped possum (*Dactylopsila trivirgata*) lives in a small pocket of Australian habitat, right up in the Cape York Peninsula. It's known to hide out in the rainforests and eucalypt woodlands, nestled on a bed of leaves in a cosy tree hollow. Sometimes it can be seen weaving among the branches as it snacks on flowers, fruits, and beetle larvae.

The striped possum is very rarely seen in Australia, but not hard to find in New Guinea, where it's far more widespread. It's also found on a few of the tiny nearby islands, including Serui and Waigeo.

While Australia shares the striped possum with New Guinea, New Guinea has three more species in the genus that never made it here: the great-tailed triok (*Dactylopsila megalura*), the long-fingered triok (*Dactylopsila palpator*), and Tate's triok (*Dactylopsila tatei*).

A close relative of the sugar glider, the striped possum looks like a black and white squirrel and has a prehensile tail like a monkey, but it's more like a marsupial woodpecker.

It feeds on wood-boring insect larvae that lurk beneath the surface of rotting branches, and licks sap, gum, and honey that seeps down onto the bark.

Just like an aye-aye, it uses its creepy elongated fourth finger to tap on the surface of a tree, listening for signs of inhabitants.

When it's located its target, it will bite a hole in the bark with its chisel-like teeth, and use the specialised hooked nail on the end of its long finger to skewer or scoop out the larvae. It also makes a meal out of ant and termite eggs using this method.

A shy and solitary species, very little is known about how the striped possum breeds. But the female has two teats in its pouch, which suggests she can raise two young at a time.

The good news is the striped possum, while extremely elusive and rare in Australia, is not considered to be a threatened species. But its numbers are in decline due to habitat loss. In New Guinea, they are also hunted for their pelts and meat, and in Australia, amethystine pythons – Australia's longest python (see page 101) – can be a real problem for the species.

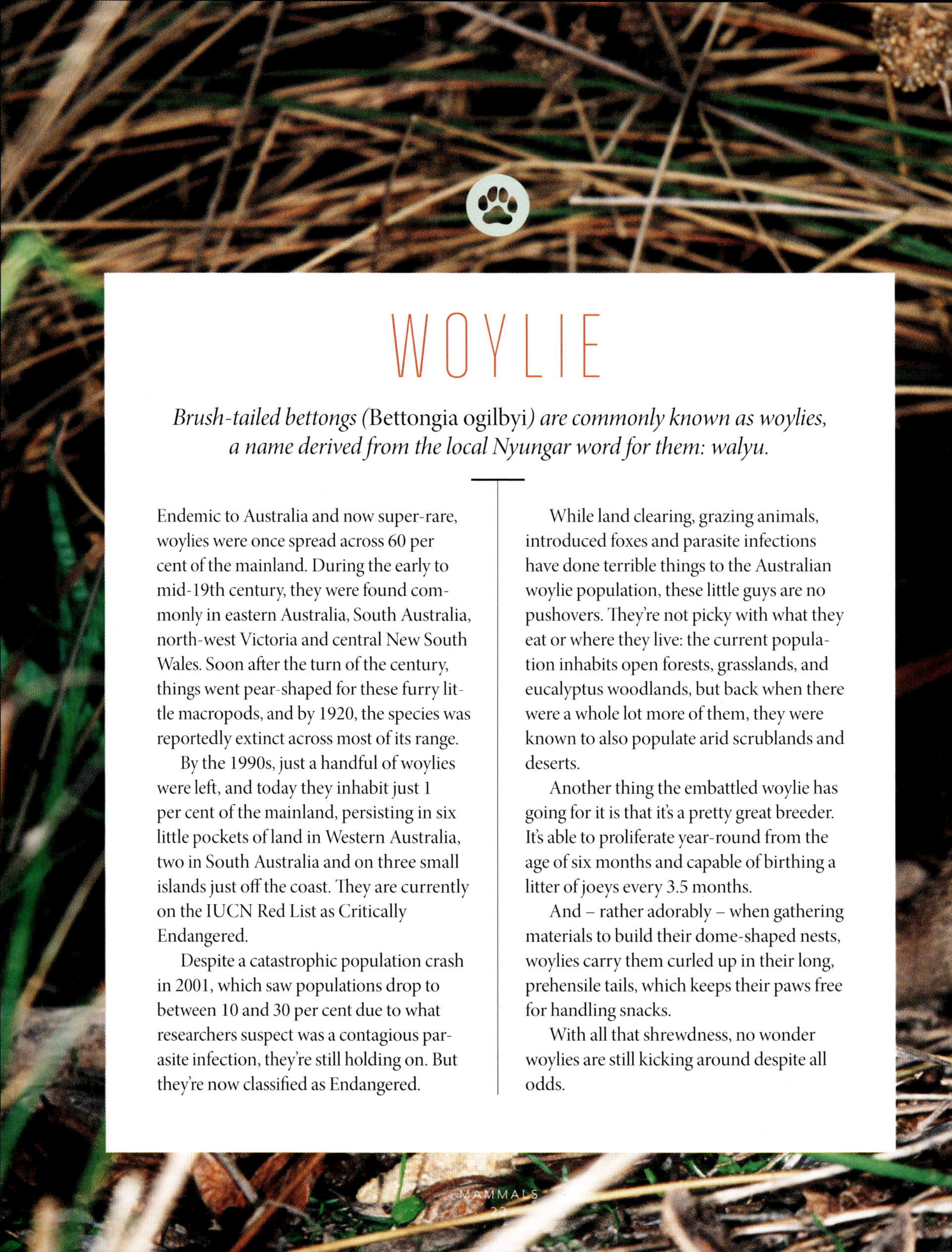

WOYLIE

*Brush-tailed bettongs (*Bettongia ogilbyi*) are commonly known as woylies, a name derived from the local Nyungar word for them: walyu.*

Endemic to Australia and now super-rare, woylies were once spread across 60 per cent of the mainland. During the early to mid-19th century, they were found commonly in eastern Australia, South Australia, north-west Victoria and central New South Wales. Soon after the turn of the century, things went pear-shaped for these furry little macropods, and by 1920, the species was reportedly extinct across most of its range.

By the 1990s, just a handful of woylies were left, and today they inhabit just 1 per cent of the mainland, persisting in six little pockets of land in Western Australia, two in South Australia and on three small islands just off the coast. They are currently on the IUCN Red List as Critically Endangered.

Despite a catastrophic population crash in 2001, which saw populations drop to between 10 and 30 per cent due to what researchers suspect was a contagious parasite infection, they're still holding on. But they're now classified as Endangered.

While land clearing, grazing animals, introduced foxes and parasite infections have done terrible things to the Australian woylie population, these little guys are no pushovers. They're not picky with what they eat or where they live: the current population inhabits open forests, grasslands, and eucalyptus woodlands, but back when there were a whole lot more of them, they were known to also populate arid scrublands and deserts.

Another thing the embattled woylie has going for it is that it's a pretty great breeder. It's able to proliferate year-round from the age of six months and capable of birthing a litter of joeys every 3.5 months.

And – rather adorably – when gathering materials to build their dome-shaped nests, woylies carry them curled up in their long, prehensile tails, which keeps their paws free for handling snacks.

With all that shrewdness, no wonder woylies are still kicking around despite all odds.

Woylie

ANTILOPINE WALLAROO

As the only large species of macropod in the world that lives entirely in the tropics, this wallaroo is a unique creature with some truly bizarre behaviours.

This mammal has kept scientists guessing about its diet, biology and breeding habits, but what we do know is fascinating. Like the red and eastern grey kangaroos, the antilopine wallaroo is highly sexually dimorphic, which means it's really easy to tell the difference between a male and female just by looking at them.

Adult males have a reddish tan coat that coordinates beautifully with their cream-coloured underbelly and they have black-tipped paws and hind feet. They also display a distinct swelling of the nose above the nostrils, which is likely an adaptation to help them pant in high temperatures.

Male antilopine wallaroos are thought to grow to as much as 70kg, which is 2.5 times the weight of a Dalmatian.

Females are typically three to four times smaller than the males, growing to just 15–30kg as adults. Their coat is grey on the head and shoulders and brownish tan on the back and hindquarters.

And while they have the same black tips on their paws and hind feet and the cream undersides, they have distinct white tips on the back of their ears.

Antilopine wallaroos are very social animals and display an unusual behaviour called sexual segregation, which sees them break off into single-sex groups during breeding season.

It's not known exactly why they do this, but it could have something to do with the unique diet the females require when they're reproducing, or it could be an anti-predator strategy.

There are three species of wallaroos in Australia (wallaroos are typically larger than wallabies but smaller than kangaroos) and the black and common wallaroos are solitary creatures.

Antilopine wallaroo

ANTILOPINE WALLAROO

The fact that the antilopine wallaroo likes to group up makes it weirdly roo-like, which is why you might see scientists refer to it as an antilopine kangaroo.

These are truly beautiful animals, but according to those who know them best, they're awkward, blundering and are quick to develop a chronic sucking behaviour that can seriously stunt their growth.

"In my experience, antilopine wallaroo joeys are incredibly clumsy (a few could even be described as dumb) and therefore vulnerable to self-injury and a number of other problems," Sarah Hirst, biodiversity liaison officer at the Territory Wildlife Park in Darwin, reported at a National Wildlife Rehabilitation Conference.

"Some of the more unbelievable happenings include joeys that routinely chase cars, some for several kilometres; and two separate incidents of joeys that jumped into the coals of a campfire and singed their feet."

Hirst has also noticed that, just like many humans, antilopine wallaroo joeys like to suck things if they're hungry, or even as a type of 'security blanket'.

The behaviour emerges in the pouch and can be almost impossible to break if it's retained into adulthood, leading to some rather poignant displays.

"The sight of an adult female sucking on the ear of her pouch joey is truly sad! Another case is that of a three-year-old male that still sucks his toe. He is now only about half the size of his counterparts and is also smaller than males a year younger", says Hirst.

Antilopine wallaroo

MULGARA

The mulgara is a close relation of the Tasmanian devil and the quoll. With a length of 20cm – half of which is tail – it is much smaller than its carnivorous cousins, but it's just as vicious and toothy.

While this desert-dwelling predator usually subsists on a diet of insects and spiders, if it comes across a lizard, mouse, or newly hatched snake, it'll happily sink its fangs in and make a delicious meal out of it.

There are two known species of mulgara: the brush-tailed mulgara (*Dasycercus blythi*) and the crest-tailed mulgara (*D. cristicauda*). You can pretty easily figure out what species a particular mulgara belongs to by checking how fluffy its tail is and how many teeth and nipples it has, but it's their range that really sets them apart.

The brush-tailed mulgara enjoys an extensive range running right through the middle of Australia, where it thrives in the arid spinifex grasslands of South Australia, Western Australia and the Northern Territory. The crest-tailed mulgara, on the other hand, is found only in a tiny pocket of the southern Simpson Desert in Queensland.

Like the lovely little western pygmy possum, the mulgara whiles away much of its life in a state of torpor. This energy-preserving period of reduced physiological activity can last anywhere from three to 12 hours a day, which is perfect for pregnant mulgara mums that don't want to deal with being pregnant all the time. It's tough carrying around all that extra weight!

A 1994 study on crest-tailed mulgaras led by zoologist Fritz Geiser from the Department of Zoology at the University of New England in Armidale found that during the species' reproductive season – which runs from June to December – 75 per cent of the pregnant females fell frequently into the deep sleep of torpor.

Not the males though – less than half of them were found to regularly display torpor during this time, because…well, their genes aren't going to pass themselves on!

SOUTHERN MARSUPIAL MOLE

This preposterous mole seems to have dressed in the dark. It is an eyeless creature with comically large claws, a barely-there tail and a luxurious golden coat.

We actually really love this plucky little weirdo. Found in the western central Australian desert, the southern marsupial mole (*Notoryctes typhlops*) is one of Australia's most elusive creatures. Also known as the itjaritjari, it's an important ancestral creature for the Anangu people of the Red Centre and features in the Dreaming of several more Aboriginal groups.

Stretching just 10cm long and weighing between 40 and 70g, this evolutionary marvel is built for burrowing. Southern marsupial moles spend almost no time on the surface, which means even those living alongside the species rarely see it.

So little is known about these shy, solitary creatures that scientists aren't entirely sure how they find each other atop or below the sandy dunes. How do they court each other? What happens when they mate? Are the males territorial? No one's been able to keep one alive in captivity long enough to breed and we're still searching for a wild population to observe.

What's particularly strange about the southern marsupial mole is that, unlike other subterranean creatures like the naked mole rat and the prairie dog, it doesn't appear to construct any burrows. Instead it just 'swims' through the sand like a fish through the ocean, backfilling its tunnels as it goes and breathing the air between each grain of sand.

As Dr Joe Benshemesh from the University of Melbourne and the Threatened Species Recovery Hub reports, fossil evidence suggests that the Australian marsupial moles likely split from the marsupial family tree some 64 million years ago. Since then, they've been evolving some weird features, such as a shield-like nose, the rear-opening pouch and complete lack of optic nerves.

Southern marsupial mole

Blanket octopus

FISH AND SEA CREATURES

HAIRY SQUAT LOBSTER

Somehow this miniature hairy squat lobster ended up with a near-fluorescent pinkish-purple hue, little orange eyes and a thick golden fleece all over. And at just a few millimetres long, it's as compact as it is adorable.

Seriously, just stop and consider how small that actually is. This bewhiskered 'fairy crab' is smaller than your fingernail, which makes it the tiniest teddy bear of the ocean. That explains how they got their nickname.

Found off the coast of Western Australia, Indonesia, the Philippines and Japan, the hairy squat lobster (*Lauriea siagiani*) makes its home on giant barrel sponges (*Xestospongia testudinaria*), where it sits and catches ocean detritus in its 'fur.' It then uses its smaller, 'un-hairy' legs to collect the particles and eat them.

Hairy squat lobsters also use their tiny stature for mischief, stealing food from anemones – a risky affair, considering anemones wield toxins and are aggressively defended by their clownfish residents. But if you're small enough, maybe no one notices?

Despite its name, *L. siagiani* is not really a lobster, but instead belongs to a group of crabs called the Anomurans. These crabs are decapod crustaceans, with ten legs. The two legs in the rear tend to be very small and mostly hidden. The hairy squat lobster uses them to clean itself and collect its food.

Like their close relatives, the hermit crabs, squat lobsters are missing a shell, which means they have to hide in the nooks and crannies of sponges to protect their soft bodies from harm.

They also instinctually wrap their tails up under themselves for protection, which gives them their distinctly squat appearance.

The other famous squat lobster is the yeti crab, with its comically elongated, golden furry arms – a species that is thought to grow its own food on its body.

HARLEQUIN SHRIMP

*While most shrimp and prawn species are known as scavengers and bottom feeders, happy to consume almost anything they can get their pincers on, the harlequin shrimp (*Hymenocera picta*) has an unusually picky taste.*

Found throughout the tropical Indian and Pacific oceans, including off the north coast of Australia, the harlequin shrimp feeds exclusively on starfish. And unlike its flamboyant exterior, how it goes about this is anything but pretty.

When a harlequin shrimp comes across a slow-moving starfish, it will try to flip it over to immobilise it. Then it will eat the starfish alive, starting with the softest parts, such as the feet. Even if the starfish sheds one of its own limbs in a bid to escape, this only weakens it for the next encounter.

And because harlequin shrimp form monogamous pair bonds, those unfortunate starfish don't just have to contend with one harlequin shrimp – they get the double act.

When looking up this amazing species, you might also come across the species, *Hymenocera elegans*.

The classification of the harlequin shrimp is confusing, because while there is officially only one species in the genus *Hymenocera*, which is *Hymenocera picta*, some researchers think there are actually two species: *H. picta* from the central and east Pacific and *H. elegans* from the Indian Ocean and west Pacific.

While harlequin shrimp might be bad news for starfish, that could make them good news for coral reefs.

The venomous Crown-of-thorns starfish (*Acanthaster planci*) are notorious for invading coral reefs in huge numbers and destroying them, and harlequin shrimp are important predators of these voracious pests.

Like the dusky batfish (page 39), which can clear reefs of great swathes of seaweed and algae, the challenge for scientists is figuring out just how to deploy them.

Harlequin shrimp

Dusky batfish

DUSKY BATFISH

*A member of the spadefish family (Ephippidae), named for their wide, flat and often triangular shape, the dusky batfish (*Platax pinnatus*) is one of five known species of batfish in the world.*

The dusky batfish is like a phoenix rising from the ashes. Girded by a golden halo that lights up its otherwise ashy body, the dusky batfish looks otherworldly as it flutters through the ocean.

Sadly, this striking gold and black visage fades once the dusky batfish grows into its mature form. As an adult, the dusky batfish is more silver than black, and its gold turns to yellow.

Found throughout the tropical West Pacific, the dusky batfish has a wide range in Australian waters, from south-west Western Australia to up around the northern coast and down near the central coast of New South Wales.

Its unusual appearance as a juvenile is a beautiful example of mimicry. It looks just like the *Pseudoceros* flatworm – a strongly toxic species that presents an unattractive prospect for any would-be predators.

The dusky batfish doesn't just look spectacular – it's a helpful little fish, too.

In 2005, scientists from James Cook University in Queensland ran an experiment to see which reef fish species could clear the Great Barrier Reef of the great swathes of seaweed and algae that crowds out the coral.

While the parrotfish and surgeonfish struggled to put a dent in the amount of seaweed they encountered, the dusky batfish halved the amount of seaweed within five days and cleared it completely in eight weeks.

"Worldwide, coral reefs are in decline," one of the team, Dave Bellwood, said at the time. "Commonly this takes the form of the coral being smothered by weedy growth, a transition known as a phase-shift, which is very hard, if not impossible, to reverse."

The bad news is that dusky batfish are quite rare around the Great Barrier Reef, so researchers are continuing the search for species that can assist in efforts to bolster coral ecosystems.

SPOTTED GARDEN EEL

These frowny eels look like they're ready to scream at the kids to get off their lawn before retreating underground to write a complaint letter.

Meet the spotted garden eel (*Heteroconger hassi*), a strange little species with a body shaped like a rubber pencil and personality for days.

Found throughout the tropical and subtropical waters of the Indo-Pacific, spotted garden eels have been recorded off the coast of Australia in the Timor Sea, the Coral Sea and the Great Barrier Reef.

While they might look like belligerent loners, spotted garden eels are social animals, known for congregating in large, tightly knit colonies.

They build their burrows mere centimetres away from each other, and when they're poking their heads up and bobbing in unison, they really do look like garden foliage, swaying in the breeze.

Spotted garden eels use their large (relative to body size) mouths to catch plankton as it floats past. If they're not feeding, they'll be hiding in their burrows, which they carve out of the seafloor using their stiff, muscular tail.

They have specialised glands in their skin that secrete a mucus that hardens like cement. The mucus not only fortifies the eels' burrows so they don't cave in, but is also used to create a temporary seal at the entrance, blocking any would-be intruders from getting in.

Unfortunately for the spotted garden eels, certain predators have figured out other ways to get to them. Snake eels will dig into the sand and attack the spotted garden eel from below. Triggerfish will chase the eels into their burrows, then 'dive-bomb' the seafloor to force them out.

Life in the ocean can be tough for these eels. No wonder they scowl at everyone.

BUBBLE ALGAE

You know what's weird? Looking at something large enough to hold in your hand and knowing it's made up of a single, solitary cell.

We're used to thinking about cells as microscopic building blocks of life. We learn in high school biology that there are simple, single-celled organisms, but we're used to them being microscopic and impossible to perceive with the naked eye.

But then there's bubble algae (*Ventricaria ventricosa*, formerly *Valonia ventricosa*), a species that is neither plant, nor animal, and at up to 9cm in diameter, is one of the largest single-celled organisms on Earth.

Found in tropical and subtropical ocean waters across the globe, including off the coast of Australia, bubble algae sit among coral rubble and mangroves, their unusual sheen making them appear like giant pearls below the surface.

They come in shades of green, from bright emerald to a dark moss, but can appear silver, teal or black through the water. Their uncanny spherical shape has earned them the nickname 'sailor's eyeballs'.

So how did something that only consists of a single cell grow to be so large?

Well, bubble algae are known as coenocytic organisms, which means they have many cell nuclei, but they're not separated by cell walls. This occurs when there is repeated division of the nucleus in the original cell, but not of the cytoplasm, which is the gelatinous liquid that fills the inside of a cell.

This means that, technically, bubble algae only have one cell, but that single cell contains the parts of multiple cells, which allows the strange species to grow to a macroscopic size.

Bubble algae can grow very rapidly, which makes its invasive potential a real worry. They are a notorious scourge among aquarium enthusiasts, who say that if you pop one, you'll end up with many more.

FLAMBOYANT CUTTLEFISH

You know those dogs whose back legs don't work but they fly around in a wheelchair, pulling themselves along with their two front legs? That's pretty much how the flamboyant cuttlefish gets about.

If you assumed cuttlefish swim through the ocean like squids and octopuses, you'd be mostly right, but on this (very) rare occasion, we find a species that prefers to traverse the seafloor, using its front fins to pull itself over bumps and crevices.

Scientists think the flamboyant cuttlefish has opted to walk rather than swim because its cuttlebone, which is that flat, chalky shell you'll often find washed up on the beach, is smaller relative to its body size than in other cuttlefish species.

The cuttlebone, which is filled with gas when the cuttlefish is alive, helps the cuttlefish remain afloat, but in this case, not for very long: "They flap up into the water, settle down again, and go back to walking along the seafloor," Dr Mark Norman, a marine biologist from the University of Melbourne and Museum Victoria told NPR. This walking behaviour is just one of the strange things about the flamboyant cuttlefish.

The beautiful species is found all around the northern coast of Australia, from Shark Bay in Western Australia across to Mooloolaba in Queensland, as well as in southern New Guinea and islands in the Philippines, Indonesia and Malaysia.

Described by the Monterey Bay Aquarium in California as a "perpetual colour machine", the flamboyant cuttlefish creates moving colour displays on its skin, ranging yellow, red, maroon, and pink, to white, brown and black. It does this by manipulating the chromatophores – pigmented, light-reflecting cells – in its skin.

Research by Dr Norman and his team back in 2007 suggested that the flamboyant cuttlefish was highly venomous but more recent research has failed to support this. So perhaps the flamboyant cuttlefish, with its bright warning colours, is just bluffing.

Flamboyant cuttlefish

Banded archerfish

BANDED ARCHERFISH

This very special fish practises social distancing even when hunting.

Not satisfied feeding on aquatic prey alone, the banded archerfish shoots a highly targeted jet of water at insects on overhanging branches to knock them into the water.

These 'spit missiles' can hit targets up to 3 metres away, and once the insect hits the water, it takes less than a split second for the archerfish to reach it and gobble it up.

At home in Australia's brackish mangrove estuaries, ranging from north-western Western Australia to northern Queensland, the banded archerfish (*Toxotes jaculatrix*) is also found throughout the Indo-Pacific, including in Thailand, China and Papua New Guinea.

Its name, *Toxotes*, borrows from the Ancient Greek word for "archer" and refers to Sagittarius, the bow-wielding centaur and ninth astrological sign.

Banded archerfish can grow to be quite large – they're typically about 20cm long, but the biggest specimens stretch to 30cm. They sport distinctive black stripes along their silvery sides and their backs have a subtle green or brown hue.

Banded archerfish are omnivores which means they'll eat pretty much anything they can get their… fins on. Edible foliage, small crustaceans and fish are all fair game in the water. Above the water, they'll target insects such as flies and crickets.

Their water jets don't just cover an impressive distance. Banded archerfish have specialised mouths and jaws that allow them to be completely in control of the force, the angle and the speed of their spit.

Like expert marksmen, they can even map out their trajectory to hit a moving target, and scientists have observed how they account for the light refraction that occurs underwater as they judge their timing. This fish can also leap up out of the water to heights of up to 2.5 body lengths to snare unsuspecting prey.

BLANKET OCTOPUS

Honestly, how are any of us supposed to get anything done when there are animals like this cruising about in the ocean like it's no big deal?

Let's take a moment to truly appreciate how bonkers this creature is. The blanket octopus (*Tremoctopus violaceus*) is a bizarre, impressive feat of evolution that might actually be even weirder than it looks.

How is that even possible, you might ask? Go with me here because there's a lot to this "beautiful scarf of living flesh", as one scientist put it.

First thing's first: the most notable thing about the blanket octopus is its sexual dimorphism.

Sexual dimorphism can manifest in many different ways. For example, only the male pink robin is pink. Size is another common one: New Zealand's katipo females are larger than the males (like our redback spiders).

This is also the case for the blanket octopus, but it just so happens to exhibit one of the most extreme cases of sexual size-dimorphism known to science.

The females, which can stretch up to 2m long and 10kg in weight, grow to be 40,000 times heavier and 100 times larger than the males, which are – wait for it – 2.4cm. "Imagine a female the size of a person and the male a size of a walnut," evolutionary biologist Tom Tregenza said, when the first ever sighting of a live male blanket octopus was reported from the Great Barrier Reef.

The females use their fleshy cape to make themselves appear intimidating to potential predators.

These capes are detachable in case of emergencies. The cape conceals the female's tentacles, which are also detachable. In fact, when a male finds a female in the open ocean, he will fill one of his tentacles up with sperm like a sock, tear it off, present it to her, and then float away to die.

Blanket octopuses are found all over the world, so if you're a diver, keep an eye out for these otherworldly beauties.

SHAME-FACED CRAB

It may look like it's hiding its face out of embarrassment, but this crab has everything to be proud of.

I don't think I've ever loved another crab as much as I love this crab right now. It's so embarrassed, it can barely even look at us. It's so ashamed, it has to cover its face with its two humongous front pincers. Don't feel bad, shame-faced crab, we don't care what you did; you're just a crab.

Found at depths of up to 50m below the surface of the Indo-Pacific, these crabs range as far as Madagascar, Japan and throughout Indonesia, Papua New Guinea and New Caledonia. They've also been spotted off the coast of Western Australia

Despite their looks and their funny little name, shame-faced crabs (*Calappa calappa*) are no victims. These highly armoured creatures are like walking tanks, their 15cm-long, burnt-caramel coloured carapace acting as the perfect cover from predators until they have a chance to bury themselves right into the sand.

Shame-faced crabs keep themselves hidden during the day. At night they turn to hunting, targeting little molluscs such as clams, oysters and sea snails. While hard-shelled prey like these present a challenge to many would-be predators, the shame-faced crab has evolved to deal with them expertly.

Of its two huge, meaty pincers, the right one is perfect for cracking into its prey's tough outer shell. It's equipped with a single, curved tooth that works with the flat surface of the pincer just like a can-opener to cut into its prey. Then the left pincer, which is longer, smaller and sharper, takes over to extract the flesh from inside.

Being elegant about how you eat your dinner is nothing to be ashamed about, shame-faced crab. The rest of the ocean is filled with barbaric rubes that wouldn't know a utensil if it landed on them. Chin up, little guy!

Leichhardt's grasshopper

INVERTEBRATES

BLUE-BANDED BEE

*This has to be one of the prettiest bees in the world. Named for the beautiful turquoise bands that run across its abdomen, the blue-banded bee (*Amegilla cingulate*) sports lush golden and white fluff and enormous green eyes.*

Males blue-banded bees can be distinguished from females by the number of blue bands they display – they have five while the females have just four. Adult blue-banded bees typically grow to 10–12mm.

The species is found all over Australia, except in Tasmania and the Northern Territory. It's also native to Papua New Guinea, Indonesia, East Timor, Malaysia and India, so it enjoys a pretty healthy range, spreading from urban areas to open fields and dense, tropical forests.

It's rumoured that they're attracted to blue and purple flowers, perhaps because they can blend into their surroundings when collecting pollen, but this has yet to be proven.

Blue-banded bees are known to frequent lavender plants, however, and according to the Australian Museum, they appear to be attracted to people in blue clothing. But it's okay, because these bees are not aggressive, and don't move around in intimidating swarms like other species. They live solitary lives in little burrows in the soil or the crevices of rocks.

Blue-banded bees are one of a few native Australian bee species that perform ▶

Blue-banded bee

BLUE-BANDED BEE

a particular type of pollination known as 'buzz pollination'. Also referred to as sonication, this type of pollination is really useful on crops such as tomatoes, blueberries, cranberries, kiwi fruit, eggplants and chilies, but the well-known and very common western honeybee (*Apis mellifera*) is incapable of performing this process. For this reason, the blue-banded bee is extremely valuable to Australian farmers.

A flower's stamen is its pollen-producing reproductive organ. Attached to the stamen is the anther, which is a one or two-lobed formation that holds onto the pollen. In some plants, the pollen is held so firmly by the anthers that it needs a little extra help breaking free, which is where solitary bees like the blue-banded bee come in handy.

These bees will grab onto the flower, and shake their entire bodies rapidly, causing both the flower and its anthers to vibrate. This shaking movement causes the pollen to be dislodged from the anther, and then be collected by the bee.

According to the University of Nevada's Leonard Lab in the US, about 8 per cent of the world's plants need to be buzz pollinated in order to reproduce.

Blue-banded bee

Processionary caterpillar

PROCESSIONARY CATERPILLAR

If you're looking for something dangerously stupid, look no further than the delightfully contradictory processionary caterpillar.

In one moment, these congregating larvae can be found marching in a never-ending circle of mindless confusion. In another, they're killing unborn foals in the womb.

The processionary caterpillar (*Ochrogaster lunifer*) has at least 2 million finely barbed hairs, which could inflict a nasty case of hives – or worse – if you happen to touch one.

Found throughout coastal and inland Australia, they feed on acacias and beefwood, as well as eucalypts. They display a fascinating behaviour that helps them find new sources of food.

Once these caterpillars have stripped a plant of its edible foliage, they will wander off to find more, leaving a silken trail behind them.

Processionary caterpillars instinctually follow these silken trails, so it's not uncommon to see groups of up to 200 of them moving head-to-tail in a tightly formed line, seeking a fresh supply of food.

This behaviour has its downsides. If one caterpillar finds a silken trail left by another, they will follow each other around in an endless circle. And if a whole group of processionary caterpillars do this, they will end up in one big, confused mess.

The real bummer about these insects is the fact that, even if pets and farm animals are cluey enough to keep away from the caterpillars themselves, their loose hairs can cause a world of pain if accidentally ingested.

As Tim Low reported for Australian Geographic in 2017, these caterpillars are a particular problem for horse breeders. "Should a pregnant mare ingest some of these hairs when they fall on grass, they can penetrate the intestinal wall, allowing bacteria into the bloodstream and infection of the placenta," he explains. "The caterpillar hairs are a dire problem in the Hunter Valley, where they've caused hundreds of thoroughbred foals to be aborted."

FISHING SPIDER

The idea that water provides no escape from spiders is not great. They lurk in our bushlands, gardens, back sheds, bedrooms – we've already got snakes to think about in the water, do we have to worry about spiders in there, too?

The short answer is no, because fishing spiders are far more interested in fish than they are in humans. The more complicated answer is yes, because my irrational fear cares not for your logic.

Fishing spiders are found on every continent on Earth, bar Antarctica. Most of them belong to the *Dolomedes* and *Nilus* genera, known for their formidable size. The leg-span of a female *Dolomedes*, for example, can stretch to as much as 9cm long.

Fishing spiders spend most of their time at the water's edge, camouflaging against rocks or foliage alongside shallow freshwater streams, rivers, lakes, ponds and swamps. *Dolomedes*, in particular, are exceptional at swimming, diving and walking on the water's surface, which helps them to snatch and eat nutrient-rich meals of local fish species. No prizes for guessing where the biggest fish caught by a spider was from, because yep, it's Australia. The largest fish on record captured by a fishing spider was a goldfish in a garden pond in Sydney measuring 9cm in length and weighing approximately 10g.

I'll end this swimming spider nightmare with a cute fact: fishing spiders belong to the family Pisauridae, a.k.a. nursery web spiders.

When the mother spider senses that her eggs are about to hatch, she goes ahead and builds a nursery 'tent', into which she manoeuvres her ripening egg sac. She then stands guard atop the tent tent, hugging it close to her body and gripping onto it with her mandibles. It's not an image this arachnophobe enjoys, but I appreciate the ingenuity.

HEAD-STACKING CATERPILLAR

Someone needs to give this guy a part in the next Mad Max *film, because that is some post-apocalyptic warlord fashion, if ever I've seen it.*

The head-stacking caterpillar is the larvae of the gumleaf skeletoniser moth (*Uraba lugens*). Even its name is metal. Like many animals that are contained within an exoskeleton, this caterpillar, known for chewing around the veins of eucalypt leaves, leaving 'skeletons' behind, must moult regularly in order to grow.

While some species, like yabbies, eat their moulted exoskeleton to take in all those good nutrients, this caterpillar delicately sheds its old head and hangs on to it. The more heads it sheds as it grows, the more elaborate its display becomes.

A recent study led by Petah Low, a biologist at the University of Sydney, suggests that the stacking head display serves as a protective deterrent against the caterpillar's natural predators, such as birds, reptiles, parasitic wasps and flies, stinkbugs, jumping spiders and lacewings.

Low and her colleagues set up experiments to observe the survival rates of caterpillars with and without a head stack. The stacks ranged from one to four heads high.

They found that the head stack can act as a false target for predators ('decoy mechanism') and as a weapon against the smaller insects ('lance mechanism').

While the head stack ultimately did not deter predators that were particularly set on making a meal out of the caterpillar, it certainly made things more difficult.

If that vertically inclined conga line of decapitated heads isn't enough of a deterrent, the hairy fibres emanating from its body can cause severe skin irritation. This a look, but don't touch, kind of caterpillar.

HAMMERHEAD WORM

Who's got no respiratory system, no circulatory system, no skeleton and a mouth for an anus? This guy.

The aptly named hammerhead worm belongs to one of the most primitive animal groups in the world – the Platyhelminthes, otherwise known as flatworms. Predatory and large, with some species capable of growing to half a metre long, hammerhead worms are a global menace, but not because they pose any threat to humans.

A number of hammerhead worm species feed exclusively on earthworms, literally tearing them apart before dissolving them in enzymes and drinking them up. And as anyone who's ever had a garden will know, earthworms are the good guys.

Native to the tropic and temperate zones of Asia and Australasia, hammerhead worms of the *Bipalium* genus have invaded almost every corner of Europe and the United States. They love anywhere that's dark, cool and moist, like piles of humus, or under rocks, logs or shrubs. While they hate anywhere dry, they can withstand brief periods of desiccation by coiling themselves into a tight ball and enveloping themselves in mucus.

Hammerhead worms glide across the ground on a single mucus-fuelled 'creeping sole', holding their half-moon-shaped heads aloft while moving them back and forth and side to side. Like an old man combing the beach with a metal detector, hammerhead worms use the chemoreceptors on the lower surface of their heads to sense earthworm mucus and other secretions on the substrate.

Once an earthworm has been located, the hungry hammerhead worm will subdue it in a coat of mucus and cut into tiny pieces. ▶

Hammerhead worm

HAMMERHEAD WORM

To feed, a hammerhead worm will extend its pharynx – or throat – out of its mouth, secrete earthworm-dissolving enzymes onto its prey and digest the softened flesh. Once the earthworm has been processed for nutrients, it will be excreted out of the same orifice it was confused by.

You might think that the perfect revenge on the earthworm-ravaging hammerhead worm is to cut it into pieces, but that's actually the worst thing you could possibly do to a hammerhead worm if you want to get rid of it.

If a hammerhead worm is cut into bits, either lengthwise or across its body, each piece will become a new, perfectly functional worm over the course of two or three weeks. During this time, the fragments will grow longer and narrower, and a flattened, spade-like head will form on one end.

Some hammerhead worms, like the Southeast Asian species *Bipalium kewense*, will even split themselves into multiple pieces deliberately as part of an asexual reproductive strategy known as fragmentation.

By sticking the tail-end of their body to the ground and pulling away from it, hammerhead worms can break off a piece of themselves that will become a younger version. These immediately mobile fragments only take a couple of weeks to reach adulthood. A single hammerhead worm can release one to two fragments for reproduction each month.

Hammerhead worm

Wrap-around spider

WRAP-AROUND SPIDER

Let's face it, Australia has some objectively intimidating spiders. But the silver lining of being on an island that's crawling with arachnids is we get the good with the bad. The fuzzy weirdos with the stone-cold killers.

I could be talking about jumping spiders, which are just lovely. But this time, I'm talking about wrap-around spiders. Something this good at hiding shouldn't be so adorable.

Wrap-around spiders belong to the *Dolophones* genus, and there are 17 species that are endemic to Australia and parts of Oceania. They can perfectly flatten themselves against the surface of a branch because of their unique body shape. With an abdomen shaped like an inverted dish, their concave underbellies can hug the curves of a tree as easily as their fuzzy legs.

Their camouflage, which is spectacular, is aided by the peculiar pattern of oval discs that run across their abdomen, giving wrap-around spiders their other nickname: leopard spiders.

Wrap-around spiders are more active at night and will spring to life in the dark to construct a vertical orb web. Before the Sun comes up, they will destroy their web and take its position on a branch once more to hide out during the day.

So yes, if you make a habit out of grabbing random tree branches in the bush of an afternoon, you are at risk of getting a shock from a *Dolophones* spider.

The good news is they belong to the Araneida family of orb-weavers, which do wield venom, but it's not particularly dangerous to humans. The larger females also only grow to around 8mm in size.

I didn't think anything could knock the fantastic mirror spider off its perch as the coolest spider, but the wrap-around spider looks so happy to just do nothing but cuddle up and chill on a branch. And that's something I can definitely get behind.

DADDY LONGLEGS

The daddy longlegs myth, debunked.

As a fixture in Australian homes, the daddy longlegs (*Pholcus phalangioides*) has earned the kind of respect that only comes from being so ubiquitous. There's no escaping them, they're always there. It's almost like we build our houses around them.

And to add to the mystique of this dogged house guest, there's that evolutionary paradox that many of us have marvelled at – the daddy longlegs is the most venomous spider of them all, but because its fangs are so small and curved (or 'uncate'), it's totally incapable of injecting its venom into humans. Except it's all a myth.

We know this for two reasons. The first is that brown recluse spiders from North America also have small, uncate fangs, and no one's questioning their ability to inflict a nasty bite on a human. The second is thanks to a horrifying experiment on MythBusters, where Adam Savage baited a daddy longlegs to bite him.

The good news is daddy longlegs venom is almost completely harmless to humans. In fact, it's even fairly weak when inflicted on mice and insects. When Savage gets bitten by a daddy longlegs, he describes a slight burning sensation that only lasts a few seconds.

So, where did the myth of the daddy longlegs' incredibly potent venom come from? It's thought that the daddy longlegs has made of a legend of itself due to its ability to crush a redback spider in a one-on-one battle. Because surely a spider that can fell one of Australia's most feared species must outrank it on the venom scale, right?

The truth is, it's not venom that gives the daddy longlegs the upper hand over the redback spider – it's those spindlylegs. The trick is that the daddy longlegs' legs are so thin, it's almost impossible for a redback to land a bite on one.

The daddy longlegs doesn't come out of this myth-busting being any less cool. It can even take down a brown snake if it wants to, because nothing beats a carefully laid silk snare.

LEICHHARDT'S GRASSHOPPER

Named after Ludwig Leichhardt, a German naturalist and explorer who traversed the unforgiving terrain of central and northern Australia to complete one of the longest overland journeys in the history of Australian exploration.

In 1845 Leichhardt found this species near Deaf Adder Creek, which runs off the South Alligator River on the Northern Territory's Arnhem Land plateau. Describing its "bright brick colour dotted with blue", Leichhardt reported finding the grasshopper here in great numbers, but between this sighting and the early 1900s, only two more specimens were discovered, each many years apart. And then, just like its namesake, the species disappeared without a trace.

Seventy years passed before Leichhardt's grasshopper was officially 'rediscovered' by Principal Research Scientist of CSIRO's Division of Wildlife Research, J. H. Calaby. He returned to South Alligator River in 1971, now protected within Kakadu National Park, spotting a single male nymph on a sandstone slope. While Calaby's find meant that the species was not extinct after all, it remains particularly rare and little studied, with just a few fragile populations sustained by three native species of flowering shrub within the Kakadu and Keep River National Parks.

Bright colouring in insects usually signifies some level of toxicity, as does this species' tendency to spew a brownish liquid when agitated, but chemical analysis has turned up little evidence that these grasshoppers are harbouring any toxic compounds. The species has no known vertebrate predators, which suggests that, rather than being toxic, the brownish spew is simply meant to taste awful. By feeding exclusively on bitter-tasting plants, the species ensures it's a beautiful but terrible meal.

Victoria's riflebird

BIRDS

APOSTLEBIRD

Okay everyone, you can stop looking – I've found my new favourite bird.

They feed together, gather nesting materials together, cross the road responsibly together, and hang out in the trees together looking super snug.

With a wide range throughout eastern Australia and a single, isolated population in the Elliott and Katherine regions of the Northern Territory, the apostlebird (*Struthidea cinerea*) belongs to the mudnester family (Corcoracidae), of which there is but one other member. And together with the white-winged chough (*Corcorax melanorhamphos*) – a blackish bird with hellish, red eyes – the apostlebird maintains a pretty strange nest and an even stranger social group.

When an apostlebird builds its deep, cup-shaped nest made from dried grasses held together with mud and manure, it will share the experience with around nine other apostlebirds. This 'breeding unit' comprises a single dominant male and adult female, plus several juvenile birds from a previous breeding season that stick around to help.

These young birds will remain with their parents for an incredible 200 days. To put that into perspective, a pigeon chick typically leaves the nest after about 30 days, and even seagulls – known to shadow their parents for a long time – eventually break loose after 90 days or so.

But not apostlebirds. Social groups are so important to them, they will team up with 6 to 20 other birds, doing everything together and telling each other about it very loudly. That's how the apostlebird got its name, after the 12 biblical apostles who teamed up to follow Jesus.

Unfortunately, things get kinda ugly when either the dominant male or female dies in a mudnester family group. The remaining dominant bird must form a new family group by finding a lone adult to bond and breed with. If helper numbers are low, they will need to be strengthened, but often a newly paired male and female won't have the help required to birth new offspring. So, they go on the hunt for other birds' offspring to bolster their ranks.

Recently fledged young are kidnapped from neighbouring groups and cared for, eventually becoming helpers in their new groups. As in, the kidnapped become the kidnappers. That 'family group' thing is starting to look a whole lot more creepy.

PHEASANT COUCAL

Let's not even talk about how it looks like a medieval war bird, ready to strike some rebels, or a dinosaur that walked straight out of the Cretaceous and into the present day.

We don't need to convince you that this is one of the coolest looking birds in Australia. But there's another reason why the pheasant coucal (*Centropus phasianinus*) is worth a closer look.

Unlike most cuckoo birds, the pheasant coucal actually takes its parenting very seriously. In fact, it's the only known Australian cuckoo that lays its eggs in its own nest.

With a particular attraction to the canefields of northern Australia, and tropical and subtropical forests and mangroves, the pheasant coucal is one of the few cuckoo species that lives its entire life on the ground, feeding on large insects, frogs, lizards, eggs and whatever tiny birds and mammals it can get its talons on.

That's actually where this bird gets its name from: being large and heavyset, with black, white and golden markings along its feathers, it looks just like a pheasant, running through the undergrowth.

It even startles like a pheasant. If it feels threatened, the pheasant coucal is much more likely to run for cover, because if it opts for flight, it's going to have to settle for a really clumsy exit.

Cuckoos are notorious in the bird world for being shameless parasites. They'll lay their eggs in another bird's nest while the other bird is sitting right there. It's a brilliant tactic, because the cuckoos get some other bird to spend its time and resources rearing its offspring.

Pheasant cuckoos opted out of that particular evolutionary path and are the only cuckoos in the country to look after their own young.

They take great care in constructing their nests, using long and flexible stalks of grass and stems to weave together a kind of open-ended dome on the ground.

Once the eggs are laid under the curved structure, the parents will line the floor of the nest with fresh green leaves for added softness. The males get the brunt of the babysitting duties, with the females left free to forage as they like.

Pheasant coucal

Eastern great egret

EASTERN GREAT EGRET

Now that is an intimidatingly beautiful creature.

With a plume of delicate feathers emanating from its body like a bright white halo, this bird looks positively angelic. You'd see it posing in a tree like that and immediately want to back away, eyes averted. Maybe you'd slink back later to deposit some sort of rodent or small lizard at its feet as penance for your profound mediocrity.

Meet the eastern great egret (*Ardea alba modesta*), one of four subspecies of the great egret (*Ardea alba*).

In New Zealand, the eastern great egret is known by its Māori name, kōtuku, and has long symbolised something rare and beautiful. In the 1800s, its elegant white plumage was highly valued by Māori and European settlers, and by the early and mid-19th century, it was driven almost to extinction across the country.

The numbers are now much healthier in New Zealand, and there is a permanent population living in Waitangiroto Nature Reserve on the west coast of the South Island.

In Australia, the eastern great egret is far more common and widespread. It's found in all states and territories, including Tasmania, and some have even strayed to Lord Howe Island, Norfolk Island, and Macquarie Island, which sits about halfway between New Zealand and Antarctica.

The largest and greatest concentrations of breeding colonies in Australia are found along the coast in the Top End chunk of the Northern Territory.

That halo of peacock-like feathers is an important part of the eastern great egret's courtship rituals. Also known as nuptial plumes, they emanate from the lower back, and make a wonderful display.

The birds' bill and areas of facial skin (called lores) also change during the breeding season. Usually, they are a bright yellow colour, but during the breeding season, the bill turns mostly black and the lores become a stunning shade of green. Strangely enough, the lores change back to yellow once the egrets have finished laying their eggs. The bills will remain dark throughout the breeding season.

Another amazing feature of the eastern great egret is its neck, which stretches 1.5 times the length of its own body. That makes it the longest neck of all the white egrets and herons found in Asia and Australia.

COMB-CRESTED JACANA

The jacanas are a family of wading birds that are spread across Asia, South and Central America, and Australia.

It might look like a poor little waterbird with a horrifying number of extra legs, but the good news is that's a perfectly healthy jacana and those are its babies' legs, dangling from its chest. Which doesn't actually sound much better, but rest assured, this is all fine.

There are just eight known species of jacanas, and they're easily identifiable for having extremely long, thin toes, which allow them to walk across lily pads and other floating vegetation with ease.

While they all have those expansive feet – it's where they got the nickname Jesus birds from – and a lovely almond-shaped body, jacanas are surprisingly diverse.

In Australia we have the comb-crested jacana (*Irediparra gallinacea*), found in the freshwater wetlands of northern and eastern Australia, particularly along the coast. Its range runs from the north-eastern Kimberley in Western Australia to the Cape York Peninsula then down along the east coast to the Hunter region of NSW.

The species, which sports a distinctive red wattle on its forehead, is also found in New Guinea and Southeast Asia.

But back to the issue at hand: what's with all those feet? Jacanas are unusual birds in that the females are typically larger than the males and the males are responsible for brooding the eggs. They do it by nestling the eggs between their wings, hugging them close to their body for warmth.

Once the eggs are hatched, the chicks are born into a dangerous world – skipping across lily pads all day means they're exposed to whatever is lurking beneath. Upon sensing danger, the parent bird – usually the male – will signal to its chicks to assume position under its wings, and whisk them away, legs dangling in rather ridiculous fashion.

Comb-crested jacana

NANKEEN KESTREL

If ever the phrase, 'Kill them with cuteness,' applied, it would be given to the nankeen kestrel. This compact assassin has the sweetest face and a signature style that makes killing look easy.

Native to Australia and New Guinea, the nankeen kestrel (*Falco cenchroides*) is one of our most frequently sighted birds. You can spot them all across the country, including in big cities like Sydney and Melbourne, and occasionally in Tasmania.

Also known as the Australian kestrel, the species grows to about the size of a pigeon – one of the world's smallest kestrels. It's also one of just two types of raptor in Australia that hunt using suspension – not speed.

While hummingbirds are known for their ability to hover, beating their wings an incredible 80 times per second, larger animals find suspending themselves in the air far more difficult. For birds of prey, which must spend a considerable amount of time aloft scouting for rodents, insects and reptiles, the easiest solution is to glide around in circles before dive-bombing their target.

Not so for kestrels. These are the only birds of prey capable of hovering. Instead of beating their wings super-fast like hummingbirds, kestrels face into the wind and use its power to hold them in place as they scout.

The key to their hovering success is feathers that have evolved to be much stiffer than the feathers of other falcons, so are better able to withstand bending in the wind. Their wings also have specialised slots between the feathers, which let the air through to reduce turbulence.

They are so good at this 'wind-hovering' technique that they can keep their heads perfectly still in mid-air, and it's an absolutely beautiful thing to see.

VICTORIA'S RIFLEBIRD

*If you're looking for an absolute alpha, a force of nature that demands to be reckoned with, look no further than Victoria's riflebird (*Ptiloris victoriae*).*

This extravagant, fearless, little bird is here to be seen, and to steal your girl when you're not looking with his flamboyant and satisfyingly symmetrical display.

Endemic to the Atherton Tableland region of north-eastern Queensland, which is part of Australia's Great Dividing Range, Victoria's riflebird is known as duwuduwu to the local Yidinji people.

The species is one of just a few birds of paradise that are found in Australia, including the paradise riflebird (*Ptiloris paradiseus)* from the rainforests of NSW and central Queensland, the trumpet manucode (*Manucodia keraudrenii*), found in the Cape York Peninsula and the magnificent riflebird (*Ptilòris magnificus*), another Cape York Peninsula resident that really does live up to its name. Most of the 42 known species are in the rugged lowland forests of New Guinea.

You might not think Victoria's riflebird has much by the way of incredible colours, but these birds are like living jewels.

That black plumage is actually an incredibly deep iridescent purple and the males have flashes of brilliant turquoise running across their heads, chests, and tails.

The females are brown and nowhere near as spectacular.

The smallest riflebird, Victoria's riflebird measures between 23 and 25cm, but what it lacks in size, it makes up for in presence.

When a male is trying to get a female's attention, he will distort his body and puff out his feathers to highlight the turquoise patterning which shines through the low light of the forest. He will fan out his wings and sway, dancing until a female draws close and allows herself to be embraced and mated with. It's such a complex ritual that juveniles have been observed practicing in their nest.

Oddly enough, while the birds-of-paradise family (Paradisaeidae) is revered the world over for its otherworldly beauty, not a whole lot is known about them. It doesn't help that most of these birds are hidden away in some of the most remote and dense rainforests in the world.

Victoria's riflebird

Whistling kite

WHISTLING KITE

Is there anything cooler than a bird of prey that can spread fires using lit sticks, and is so sneaky, it can out-scavenge an ibis?

Meet the whistling kite (*Haliastur sphenurus*) – a real-life 'fire hawk' with more tricks up its sleeve than you might expect.

The most remarkable of these has been the stuff of legends for millennia – a phenomenon long-touted by the Alawa, MalakMalak, Jawoyn, and other Aboriginal peoples of northern Australia.

For thousands of years whistling kites have been intentionally spreading bushfires, by carrying lit twigs in their beaks and claws, to flush out exhausted and confused prey.

'Avian fire-spreading' has long been acknowledged by Aboriginal rangers as a real risk to controlled burning operations, as fire hawks can literally cause a blaze to jump across firebreaks.

"We're not discovering anything," Dr Mark Bonta, a geographer from Penn State University, told National Geographic, regarding his 2018 study on kites. "Most of the data that we've worked with is collaborative with Aboriginal peoples. They've known this for 40,000 years or more."

The paper was inspired by a passage in the 1962 book, *I, the Aboriginal*, by Australian journalist Douglas Lockwood and Waipuldanya (Phillip Roberts), an Alawa language group man from the Roper River country in the Northern Territory.

In it, Waipuldanya recalls his own observations of a fire hawk:

"I have seen a hawk pick up a smouldering stick in its claws and drop it in a fresh patch of dry grass half a mile away, then wait with its mates for the mad exodus of scorched and frightened rodents and reptiles. When that area was burnt out, the process was repeated elsewhere. We call these fires Jarulan."

A bird that clever surely doesn't stop there, right?

Not content to rest on its smouldering laurels, the whistling kite is also known to steal meals from other raptors, herons and even the poor old ibis, which honestly seems a bit low. Let the bin chicken have its garbage in peace.

PROVIDENCE PETREL

New Zealand gets a lot of airtime for its flightless, clumsy, and just-a-little-too-trusting parrot. But Australia has it's own version of the Kākāpō. The sweetest seabird, the providence petrel is so fearless of humans, it comes when you call.

The providence petrel (*Pterodroma solandri*) is one of Lord Howe Island's strangest species (which is saying a lot). This is a bird David Attenborough once described as, "Extraordinarily friendly towards human beings."

With just 64,000 left in the world, the species is classified as vulnerable and the only known breeding sites are at Lord Howe Island and Philip Island – a tiny island located 6km south of Norfolk Island in the Southwest Pacific.

Sadly, they used to also breed on Norfolk Island, but went extinct there in the early 1800s, and you won't be shocked why. As Guy Dutson reports for AUSTRALIAN GEOGRAPHIC:

"[HMS] *Sirius* was wrecked on Norfolk Island [in 1790] and 270 extra people joined the few first settlers. There were not enough supplies to feed them all, but they soon discovered that by lighting fires at dusk, a type of petrel would 'drop down out of the air as fast as the people can take them up and kill them'.

They took 2000 to 3000 birds every night for two months and named it the bird of providence. Not long after, the providence petrel became extinct on Norfolk Island."

These beautiful birds are still not without threats, and ironically, an even more endangered Lord Howe Island bird is its biggest predator – the Lord Howe woodhen (*Gallirallus sylvestris*), which has a taste for petrel eggs.

Providence petrels are silver-hued birds of roughly pigeon proportions and are graceful in both the sky and the sea. But call them down to the ground, and it's a whole different story. If you make loud, strange noises in the forests of Mt Gower, these seabirds will fall clumsily through the trees to the ground, their long, outstretched wings looking more gangly than graceful.

Red-capped parrot

RED-CAPPED PARROT

Easily identified by its crimson crown, neon green and yellow cheeks, and slender, elongated bill, the red-capped parrot sports a striking purple chest that is an absolute sight to see if you're lucky enough to spot one in the wild.

Just like the long-billed black cockatoo (*Calyptorhynchus baudinii*), an endangered species that has an even smaller range in southwest Western Australia, red-capped parrots are perfectly suited to feed on the seedpods of the marri eucalypts (*Corymbia calophylla*), which are native to the region, and are a staple of both species' diets.

Both birds feed mainly on the ground, using one of their feet to hold the pods like an ice-cream cone while they poke their bills inside to dislodge and retrieve the seeds.

The sole member of its genus, owing to its unique bill shape, the red-capped parrot is thought to be related to the *Psephotellus* genus of Australian parakeets. Its closest known relative is the mulga parrot (*Psephotellus varius*), a species native to southern Australia, known for its jade green and olive colouring.

The red-capped parrot's species name *spurius* refers to the fact that, while the males and females look very similar to one another (the females tend to be slightly less vibrant and have green and yellow spots on their red flanks), the juveniles look completely different. It's almost as if they're not related – they don't even have a red crown.

This confounded researchers early on in the species' discovery, and so in 1820, German naturalist Heinrich Kuhl gave them the species name, *spurius*, meaning "illegitimate" in Latin.

The genus name *Purpureicephalus*, is an amalgam that means "purple or reddish head" in Ancient Greek, which means, yes, the red-capped parrot's scientific name translates to "red-headed bastard", as Australian naturalist Ian Fraser and co-author Jeannie Gray point out in their seminal book, *Australian Bird Names*.

Unfortunate name aside, this majestic little bird is thriving. Unlike so many of our native species these days, its population actually appears to be increasing.

REPTILES AND AMPHIBIANS

Golden-tailed gecko

YELLOW-BELLIED SEA SNAKE

You know you're doing something right when you're the most widely distributed snake in the world.

Not only is the yellow-bellied sea snake the only sea snake in the world to have reached the Hawaiian Islands, but it's one of just two sea snakes to have ever made it to New Zealand. And did we mention that it's the most aquatic snake in the world, having never evolved the ability to survive on land? The yellow-bellied sea snake (*Hydrophis platurus*) is a powerhouse of a reptile that you don't want to mess with. Males can grow up to 7.2m long and the females are 8.8m long. They have a beautiful, dandelion-yellow belly, contrasting with that pitch black stripe along their back.

If these creatures look intimidating, it's because they are. Yellow-bellied sea snakes have no natural predators and are incredibly venomous, with toxin that can damage the skeletal muscles, inflict paralysis, cause acute renal failure and can be fatal if not treated.

The good news is there has never been a reported death in Australia due to a yellow-bellied sea snake bite, despite them being pretty common in our coastal waters. In fact, the species is the most common sea snake found on Perth beaches.

Fortunately, these slick predators save their venom for hunting – a defensive bite won't pack the toxic punch that an aggressive bite will, which is great for us humans and bad news for the snake's prey.

Spending their lives underwater – they eat, sleep and breed in the ocean – yellow-bellied sea snakes feed mostly on fish, immobilising them with a venomous bite.

While yellow-bellied sea snakes aren't interested in getting anywhere near a human, they sometimes beach themselves, and that's when they're at their most dangerous to us. Even a weak snake or one that appears to be dead can strike if you're trying to move it, so never approach a sea snake that's washed up on the shore – just call your local wildlife service.

CRUCIFIX FROG

This quite serious-looking frog, often mistaken for a toad, would put a frown on any predator's face with its gooey defence

Look at this fat little guy. With personality for days, the warty, ping-pong-ball-shaped crucifix frog (*Notaden bennettii*), is native to western New South Wales and south-western Queensland.

It is decorated with a striking black, red and green cross-shaped pattern that runs all the way across its bright yellow back.

These colours wouldn't do much to help the crucifix frog camouflage against the blackish flood plains it lives on – quite the opposite, they're there to make the frog stand out. The crucifix frog is one of the only species of Australian frog to employ aposematism, which is the use of bright patterning to ward off predators.

There's no evidence that the crucifix frog is poisonous like other brightly coloured species, but it does ooze a milky, sticky 'frog glue' from glands in its skin when threatened. The purpose of the glue could be twofold. If something tries to take a bite out of a crucifix frog, they'll get a mouth full of goo and know not to try that again. If these bites happen to come from something as small as flies or ants, there's a good chance they'll get stuck to the skin, thanks to the frog glue. The crucifix frog will then shed its skin and eat the whole thing, bugs and all. Which is so gross, yet so incredibly efficient.

The crucifix frog lives most of its life in underground burrows 2–3m deep, and only comes out after bouts of heavy rain. At this point, they'll go into a feeding and mating frenzy, revelling in the temporary pools formed by the downpour. Thick with bugs, worms and mosquito larvae, these pools are like an incredibly rich and nutritious soup for the frogs.

If their 'soup' ever runs out, the crucifix frog has another strategy for snagging a meal – toe waggling. This is the first Australian frog species that has been recorded wiggling its toes as a lure for prey.

Crucifix frog

Amethystine python

AMETHYSTINE PYTHON

There's something about being very big (or very small) that makes an animal automatically more charming. And it doesn't get much bigger than the amethystine python, or as Australians up in the north know it, the scrub python.

Growing anywhere from 3m to more than 8m long, the amethystine python (*Morelia amethistina*) is Australia's longest python. Regardless of whether you measure by length or by weight, it's one of the six largest snakes in the world.

Found in northern tropical Queensland, Indonesia, Western New Guinea, Papua New Guinea, and nearby islands, the amethystine python keeps mostly to dense, warm rainforests, but will occasionally venture out into the suburbs.

The good news is, being a python, this imposing reptile isn't venomous. Rather, it kills its prey by crushing it between its coiled, muscular body, and then slowly swallowing it whole.

But if there's one thing an amethystine python struggles with, it's keeping a low profile when it ventures near a home looking for food or shelter. It seems like wherever these pythons go, chaos follows.

Here's just a taste of what Australia's longest 'danger noodle' gets up to.

In 2018, Queensland electrician/ex-wildlife handler, Brydie Maro, was tasked with removing a female scrub python that had taken shelter under a family home.

As the snake was gently loaded into Maro's toolbox to be relocated, Maro noticed a very suspicious-looking bulge in the snake's otherwise slender body. As Maro declared, "Unfortunately, that's the family's cat."

In December 2017, police officers in Wujul Wujul, some 345km north of Cairns, were forced to usher a lengthy amethystine python from the middle of the road. It's estimated that this particular individual was about 5m long. At the time, the officers said there was no passing in their police vehicle until the lingering python was "good and ready to move".

Why is this mighty snake called 'amethystine'? If you look closely, and in the right light, you'll see this truly unique creature shines like an amethyst gemstone.

GOLDEN-TAILED GECKO

If you're going to get lost in someone's eyes, let it be the golden-tailed gecko's.

Like a round, red marble, with an abyssal tear down the middle like the Eye of Sauron, the golden-tailed gecko's eyes are amazing (see page 95 for a closer look).

Geckos have some of the most incredible eyes in the animal kingdom, with certain species displaying almost fantastical patterns across their irises, due to the way their blood vessels show through.

Their vision is also 350 times better than a human's, and it gets better, because while we're stuck seeing nothing but shadows at night, geckos have evolved eyes that can see in full colour by the dim light of the Moon.

But its eyes aren't what the golden-tailed gecko (*Strophurus taenicauda*) is named after. As adults, these geckos develop a golden or bright yellow stripe down their tail, which contrasts wonderfully with the rest of its black and silver body.

The species is endemic to Australia and occurs exclusively in Queensland, in the southern and central Brigalow regions. It's currently listed as Near Threatened because of severe habitat loss and feral predators.

While they struggle to fend off curious cats and dogs, geckos from the genus *Strophurus* can certainly hold their own against native threats like birds.

Strophurus geckos are just as happy up in the trees or in shrubs as they are on the ground, and when faced with birds in their lofty digs, they won't hesitate to squirt a putrid, sticky liquid from their tails straight at any would-be predators.

This brown, yellow, or orange liquid is harmless, but the smell is enough to deter most from pursuing a gecko meal – it's described by scientists as "nasty-smelling" with a "disagreeable taste", or perhaps more generously, "an unpleasant odour resembling that of crushed legume seeds".

And let's not question why or how they know, but researchers have reportedly mixed this defensive fluid with ammonia to create a highly flammable substance.

PIG-NOSED TURTLE

I feel like this is the kind of turtle I'd like to watch television with. It just looks like it would have lots of great opinions about things and would enjoy eating popcorn out of a big wooden bowl while perched on the arm of my couch.

The pig-nosed turtle (*Carettochelys insculpta*) is native to the freshwater rivers, streams and lagoons of the Northern Territory in Australia and parts of southern New Guinea.

With its large, webbed flippers, buttercup-yellow chin and wonderfully strange little snout, this charismatic mishmash of a reptile looks unlike any other freshwater turtle in the world.

The sex of an embryonic pig-nosed turtle is determined by the temperature of the ground beneath its egg. Once hatched, the average pig-nosed turtle will grow into a shell about 70cm long. It can end up weighing over 20kg.

It will use its long fleshy proboscis – or snout – sort of like a twin-snorkel, poking it up out of the water to breathe while the rest of its body remains safely submerged.

The pig-nosed turtle's paddle-shaped flippers are more like something you'd see on a marine turtle, and yet its ability to move its individual digits is more typical of a freshwater turtle species.

These features, which are seemingly at odds with each other, have led researchers to believe that their family Carettochelyidae represents an evolutionary link between freshwater and marine turtles.

Unfortunately, pig-nosed turtles are the last remaining members of this family, and thanks to their cute faces and total weirdness, they're in serious danger of disappearing too.

According to Ben Guarino, a journalist for The Dodo website, the pig-nosed turtle has unwittingly become a popular subject of black market trade and their conservation status now sits at Vulnerable.

Their unique look makes them much-sought-after pets and rumours of their supposed medicinal properties continue to flourish throughout Asia.

PERENTIE

Australia is littered with all kinds of weird and wonderful creatures. This gigantic monitor is proof that we have the best reptiles in the world.

Back in 2015, the world's smallest monitor was discovered in the Dampier Peninsula, weighing in at just 16g. At the other end of the scale, Australia's largest monitor is the mighty perentie (*Varanus giganteus*) – a legendary goanna that can match the speed of an Olympic sprinter.

At 2.5m long and up to 20kg, the perentie weighs the same as a standard poodle, and while it's not the largest monitor in the world, what it lacks (relatively) in bulk, it makes up for in style.

As the fourth largest species of lizard on Earth (after the Komodo dragon, the Asian water monitor and the crocodile monitor), the perentie is found in the arid desert areas of Western Australia, South Australia, the Northern Territory and Queensland.

According to legend, the incredible patterns that adorn the perentie's thick, leathery skin are tied to the patterns of the arguably more beautiful black-headed monitor (*Varanus tristis*), a smaller species found throughout mainland Australia and some nearby islands.

American herpetologist and evolutionary ecologist, Dr Eric Rodger, recalls the story from the Pitjantjatjara people of Central Australia of how the perentie and the black-headed monitor gave each other their spots:

"Welding the brush first, the perentie decorated its cousin handsomely. Then came the smaller monitor's turn. It started off well, the brush creating a net-like pattern from the perentie's snout to midsection, but then, frustrated by how long the job was taking, the artist picked up the paint and dumped it over the perentie's hindquarters. According to legend, the perentie continues to seek revenge and its cousin lives an arboreal [tree-based] life for fear of retribution."

Apart from its distinctive patterning, which is said to be so unique from lizard to lizard, it can be used to identify individuals like fingerprints, perenties can run at speeds of up to 40 km/hr and they can stand up on their hind legs and 'tripod' to get a better view of their surroundings.

Desert spadefoot

DESERT SPADEFOOT

With temperatures that hit 50°C in summer and freezing in winter, as well as cycles of severe drought followed by flooding, the town of Windorah in western Queensland is the kind of place that only the most specialised creatures survive.

Enter the desert spadefoot (*Notaden nichollsi*) — a comically round, stout little amphibian with stubby legs, a short snout and a sad, downturned mouth that gives it a permanent frown.

And why wouldn't it frown? It's a moisture-loving life form that's somehow found itself in one of the hottest regions on the planet.

Jannico Kelk, a wildlife and conservation photographer based in Brisbane, describes how the temporary flooding of the Windorah region is vital for the frog's breeding patterns, which involve laying clumps of over 1000 eggs at a time in the clay or water-logged sand.

"When we arrived in the red sandy dunes of Windorah, it was flooded. The normally dry ground was now enormous puddles of water interrupted only by spinifex and sand dunes.

"As night fell, desert spadefoots and a range of other frogs started to call. Desert spadefoots can stay underground for extended lengths of time to avoid drying out in the extreme desert heat, and only surface to feed and reproduce when water is available.

"We walked through the crests and troughs of dunes to see dozens of the frogs emerging from the sand. We sat down next to one that just had his eyes above ground."

These frogs are usually only seen above ground after it rains. They spend most of their lives underground, in burrows that can plunge up to a metre below the surface.

Only after flooding is the surface cool enough for the frogs to emerge, mate and deposit their eggs in the soggy sand.

And the life inside these eggs know they're on a strict deadline, too – they hatch and metamorphose from tadpole to frog in less than a month, so they can dig themselves underground before it's too late.

PICTURE CREDITS

Front cover (clockwise from top)
Jialing Cai/Shutterstock; fivespots/SS ; sacit u/Shutterstock; Possent phsycography/Shutterstock; Kristian Bell/Shutterstock; Minden Pictures/Alamy; LuFeeTheBear/Shutterstock; AGAMI Photo Agency/Alamy; Vladimir Wrangel/Shutterstock; daniilphotos/Shutterstock; Dawn photos/Shutterstock; Ken Griffiths/Shutterstock; Dirk Kotze/Shutterstock

23 Chris Watson/Shutterstock; 5 Andrea Izzotti/Alamy; 8-9 David Fleetham/Alamy; 10-11 Minden Pictures/Alamy; 13 Nature Picture Library/Alamy; 14-15 Minden Pictures/Alamy; 16 HeitiPaves/GettyImages; 18 Minden Pictures/Alamy; 21 BirdImages/GettyImages; 22-23 Martin Harvey/GettyImages; 25 Anup Shah/GettyImages; 26-27 Anup Shah/GettyImages; 28 Auscape International Pty Ltd/Alamy; 30-31 Auscape/GettyImages; 32-33 Sam Robertshaw/Shutterstock; 35 Tony Shih/Flickr; 36-37 Divelvanov/Shutterstock; 38 Gerald Robert Fischer/Shutterstock; 41 WaterFrame/Alamy; 43 WaterFrame/Alamy; 44-45 Helmut Corneli/Alamy; 46 Brian Bevan/Alamy; 49 BIOSPHOTO/Alamy; 51 Divelvanov/Shutterstock; 52-53 Nature Picture Library/Alamy; 54-55 Istiyono/Shutterstock; 56-57 aDam Wildlife/Shutterstock; 58 Chris Watson/Shutterstock; 61 Sindre Ellingsen/Alamy; 63 Tony Daley/Flickr; 64-65 one.Wilawan/Shutterstock; 66-67 BIOSPHOTO/Alamy; 71 Chris Moody/Shutterstock; 73 Manfred Gottschalk/GettyImages; 74-75 Minden Pictures/Alamy; 77 Travis k/Shutterstock; 78-79 Chris Ison/Shutterstock; 80 AlecTrusler2015/Shutterstock; 83 Sally Corte; 85 VittoriaChe/Shutterstock; 86-87 FLPA/Alamy; 88 Donovan Klein/Alamy; 91 Suzanne Long/Alamy; 93 steve sadler images/Alamy; 94-95 Steve Wilson/AustralianGeographic; 97 RobHamm/Shutterstock; 98-99 Andrew Gregory/AustralianGeographic; 100 Sibons photography/Alamy; 103 Nature Picture Library/Alamy; 105 daniilphotos/Shutterstock; 106-107 Rafael Ben-Ari/GettyImages; 108 Stefan Mokrzecki/GettyImages

Back cover Oksana Maksymova/Shutterstock

ABOUT THE AUTHOR

Bec Crew is an award-winning science communicator with a fascination for odd animal behaviours, unique adaptations, new species and the researchers that discover them. Her stories celebrate how strange yet relatable so many of the creatures living among us can be.

Based in Sydney, Australia, Bec is a Senior Editor at Nature Index, a publication of *Nature* that follows trends in scientific research. She is the former Editor in Chief at *ScienceAlert*, and was Editorial Director at STEM Matters and digital editor of COSMOS. In addition to her Creatura Blog at Australian Geographic, she has written weekly columns for *Scientific American* and *World's Best Ever*. Her first book, *Zombie Tits, Astronaut Fish and Other Weird Animals*, was published by NewSouth Press.

Bec studied classical archaeology and online journalism at the University of Sydney. When she's not scouring the internet for the next great animal to write about, she's doting on her 15-year-old tuxedo cat, Bailey, who gets away with murder like only a cat can.

Thank you to the Australian Geographic team for supporting my love of animals and giving me the opportunity to share it with others. I am particularly grateful to Angela Heathcote, Jess Teideman and Katrina O'Brien, for all of your hard work.

And a special thank you to my husband, Alex, for your rare mind and loyal heart.

CREATURA

First published in 2021

Australian Geographic
Level 7, 54 Park Street,Sydney NSW 2000
02 9136 7214

editorial@ausgeo.com.au

australiangeographic.com.au

Copyright © Australian Geographic

All rights reserved.

No part of this publication can be reproduced, stored or introduced into a retrieval system, or transmitted, in any form and by any means (electronic, mechanical, photocopying, recording or otherwise), without the prior written permission of the copyright owner and publisher of this book.

Creative Director **Mike Ellott**

Design **Harmony Southern and Mel Coggio**

Editors **Lauren Smith and Katrina O'Brien**

Subeditor **Peter Tuskan**

Australian Geographic

Managing Director **Jo Runciman**

Editor-in-Chief **Chrissie Goldrick**

The Australian Geographic Society was established to encourage the spirit of discovery and adventure, and to foster love for our natural heritage. The Society and the Australian Geographic journal sponsor scientific research and conservation, and a portion of the profits from our published products goes back into the Society. Become a member today by subscribing to the Australian Geographic journal.

Subscribe now: 1300 555 176 or australiangeographic.com.au

ISBN 978-1-92238-813-1

Printed and bound in China by Leo Paper Group

A catalogue record for this book is available from the National Library of Australia

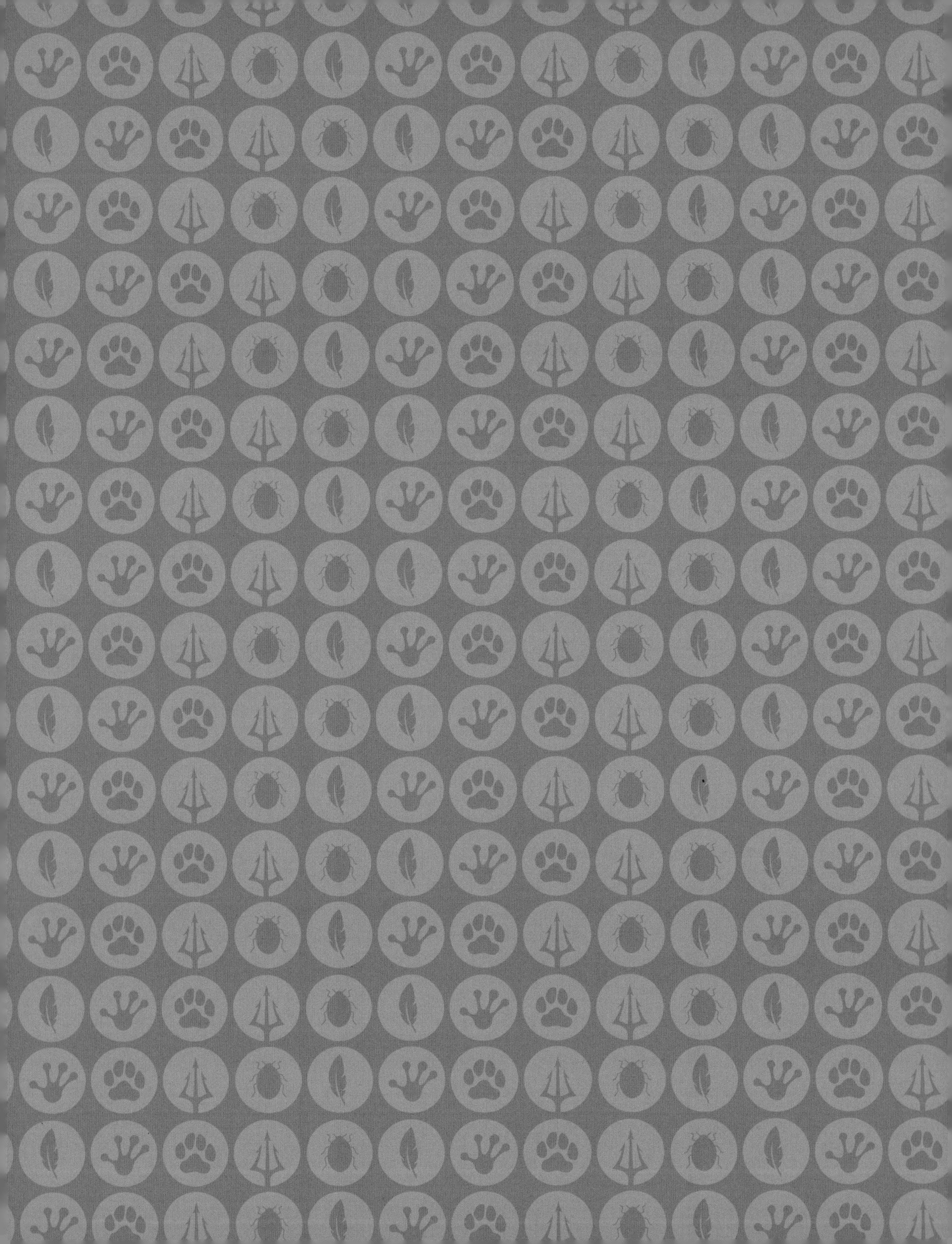